An Introduction to Subtraction

Frank Romascavage, III, PhD

fromascavage@yahoo.com

Cover design by Frank Romascavage, III

Paperback ISBN: 978-1-951816-02-5

Ebook ISBN: 978-1-951816-03-2

Introduction

The subtraction expressions from 0 – 0 to 18 – 9 as well as their respective differences are presented in the first part of this text. We will only consider expressions in this range with nonnegative differences. We will start with subtracting by 0 and will proceed in ascending numerical order to subtracting by 9. More precisely, here is the ordering of the subtraction expressions:

Subtraction by 0: 0 – 0 to 9 – 0
Subtraction by 1: 1 – 1 to 10 – 1
Subtraction by 2: 2 – 2 to 11 – 2
Subtraction by 3: 3 – 3 to 12 – 3
Subtraction by 4: 4 – 4 to 13 – 4
Subtraction by 5: 5 – 5 to 14 – 5
Subtraction by 6: 6 – 6 to 15 – 6
Subtraction by 7: 7 – 7 to 16 – 7
Subtraction by 8: 8 – 8 to 17 – 8
Subtraction by 9: 9 – 9 to 18 – 9.

On the front of the page you will be given a subtraction expression, for example 3 – 1. Next, you will have to determine the difference that results from the expression on that page. Then, you will check to see if your answer is correct by flipping to the back of the page. On the back of the page, you will find the correct answer. For example, 2 is on the back of the page that has 3 – 1 on the front, because 3 – 1 is equal to 2. If you are reading this text electronically, the answer to the subtraction expression is on the following page. For example, 2 is on the page that follows 3 – 1.

This text is recommended for anyone who wants to improve their proficiency in subtraction by a single digit. You do not have to start at 0 – 0 and proceed in order to 18 – 9 to get the most out of this text. Feel free to skip to any expressions that you want to study. For example, if you want to review 12 – 7, you do not have to solve all of the problems from 0 – 0 to 11 – 7 to get there. You can proceed directly to 12 – 7.

Ten worksheets are provided after all of the subtraction expressions and their differences are presented. Each worksheet has ten problems consisting solely of subtracting by a single digit from the expressions discussed earlier in the text. By doing the worksheets, you will have the opportunity to assess what you have learned through additional practice. Last, solutions to the worksheets will be provided at the end of the text. I hope this text will greatly increase your abilities and confidence while doing introductory subtraction by a single digit!

0—0

0

1—0

1

2—0

2

3—0

3

4 — 0

4

5—0

5

6—0

6

7 — 0

7

8 — 0

8

9—0

9

1—1

0

2 — 1

1

3—1

2

4—1

3

5 — 1

4

6 — 1

5

7 — 1

6

8—1

7

9—1

8

10—1

9

2—2

0

3 – 2

1

4—2

2

$$5 - 2$$

3

6—2

4

7－2

5

8 — 2

6

9—2

7

10－2

8

11 – 2

9

3—3

0

4—3

1

5—3

2

6—3

3

7—3

4

8—3

5

9—3

6

10—3

7

11—3

8

12—3

9

4 — 4

0

5 — 4

1

6—4

2

7—4

3

8 — 4

4

9—4

5

10—4

6

11—4

7

12—4

8

13－4

9

5 — 5

0

6—5

1

7 — 5

2

8 — 5

3

9 — 5

4

10 — 5

5

11 – 5

6

12 — 5

7

13 — 5

8

14 — 5

9

6—6

0

7 — 6

1

8—6

2

9—6

3

10—6

4

11 — 6

5

12—6

6

13 — 6

7

14—6

8

15—6

9

7 — 7

0

8—7

1

9 — 7

2

10 — 7

3

$11 - 7$

4

$$12 - 7$$

5

13 — 7

6

14 — 7

7

$$15 - 7$$

8

16 — 7

9

8 — 8

0

9—8

1

10—8

2

11—8

3

12—8

4

$$13 - 8$$

5

14－8

6

$$15 - 8$$

7

16—8

8

17 — 8

9

9—9

0

10—9

1

11—9

2

12—9

3

13 — 9

4

14—9

5

15—9

6

16—9

7

17—9

8

18—9

9

Worksheet #1

$15 - 7 =$

$4 - 3 =$

$14 - 6 =$

$3 - 1 =$

$11 - 8 =$

$9 - 2 =$

$5 - 0 =$

$13 - 4 =$

$10 - 5 =$

$12 - 9 =$

Worksheet #2

$7 - 3 =$

$8 - 0 =$

$10 - 6 =$

$5 - 5 =$

$11 - 2 =$

$9 - 8 =$

$6 - 1 =$

$10 - 4 =$

$14 - 7 =$

$15 - 9 =$

Worksheet #3

$17 - 9 =$

$11 - 5 =$

$6 - 3 =$

$5 - 4 =$

$13 - 8 =$

$12 - 7 =$

$6 - 0 =$

$9 - 1 =$

$8 - 2 =$

$9 - 6 =$

Worksheet #4

$14 - 5 =$

$11 - 6 =$

$7 - 4 =$

$4 - 2 =$

$3 - 0 =$

$8 - 1 =$

$8 - 7 =$

$13 - 9 =$

$16 - 8 =$

$11 - 3 =$

Worksheet #5

$8 - 5 =$

$7 - 2 =$

$16 - 9 =$

$12 - 8 =$

$10 - 1 =$

$0 - 0 =$

$9 - 3 =$

$11 - 4 =$

$12 - 6 =$

$11 - 7 =$

Worksheet #6

$1 - 0 =$

$9 - 5 =$

$4 - 1 =$

$11 - 9 =$

$7 - 7 =$

$9 - 4 =$

$13 - 6 =$

$5 - 2 =$

$3 - 3 =$

$14 - 8 =$

Worksheet #7

$15 - 8 =$

$5 - 3 =$

$4 - 0 =$

$2 - 2 =$

$6 - 4 =$

$12 - 5 =$

$14 - 9 =$

$10 - 7 =$

$6 - 6 =$

$5 - 1 =$

Worksheet #8

$9 - 0 =$

$4 - 4 =$

$15 - 6 =$

$7 - 1 =$

$10 - 3 =$

$16 - 7 =$

$8 - 8 =$

$7 - 5 =$

$6 - 2 =$

$18 - 9 =$

Worksheet #9

$10 - 2 =$

$7 - 6 =$

$1 - 1 =$

$10 - 9 =$

$13 - 5 =$

$7 - 0 =$

$12 - 4 =$

$10 - 8 =$

$13 - 7 =$

$12 - 3 =$

Worksheet #10

$17 - 8 =$

$9 - 7 =$

$6 - 5 =$

$2 - 0 =$

$8 - 4 =$

$2 - 1 =$

$8 - 3 =$

$3 - 2 =$

$8 - 6 =$

$9 - 9 =$

Answers to Worksheet #1

$15 - 7 = 8$

$4 - 3 = 1$

$14 - 6 = 8$

$3 - 1 = 2$

$11 - 8 = 3$

$9 - 2 = 7$

$5 - 0 = 5$

$13 - 4 = 9$

$10 - 5 = 5$

$12 - 9 = 3$

Answers to Worksheet #2

$7 - 3 = 4$

$8 - 0 = 8$

$10 - 6 = 4$

$5 - 5 = 0$

$11 - 2 = 9$

$9 - 8 = 1$

$6 - 1 = 5$

$10 - 4 = 6$

$14 - 7 = 7$

$15 - 9 = 6$

Answers to Worksheet #3

$17 - 9 = 8$

$11 - 5 = 6$

$6 - 3 = 3$

$5 - 4 = 1$

$13 - 8 = 5$

$12 - 7 = 5$

$6 - 0 = 6$

$9 - 1 = 8$

$8 - 2 = 6$

$9 - 6 = 3$

Answers to Worksheet #4

$14 - 5 = 9$

$11 - 6 = 5$

$7 - 4 = 3$

$4 - 2 = 2$

$3 - 0 = 3$

$8 - 1 = 7$

$8 - 7 = 1$

$13 - 9 = 4$

$16 - 8 = 8$

$11 - 3 = 8$

Answers to Worksheet #5

$8 - 5 = 3$

$7 - 2 = 5$

$16 - 9 = 7$

$12 - 8 = 4$

$10 - 1 = 9$

$0 - 0 = 0$

$9 - 3 = 6$

$11 - 4 = 7$

$12 - 6 = 6$

$11 - 7 = 4$

Answers to Worksheet #6

$1 - 0 = 1$

$9 - 5 = 4$

$4 - 1 = 3$

$11 - 9 = 2$

$7 - 7 = 0$

$9 - 4 = 5$

$13 - 6 = 7$

$5 - 2 = 3$

$3 - 3 = 0$

$14 - 8 = 6$

Answers to Worksheet #7

$15 - 8 = 7$

$5 - 3 = 2$

$4 - 0 = 4$

$2 - 2 = 0$

$6 - 4 = 2$

$12 - 5 = 7$

$14 - 9 = 5$

$10 - 7 = 3$

$6 - 6 = 0$

$5 - 1 = 4$

Answers to Worksheet #8

$9 - 0 = 9$

$4 - 4 = 0$

$15 - 6 = 9$

$7 - 1 = 6$

$10 - 3 = 7$

$16 - 7 = 9$

$8 - 8 = 0$

$7 - 5 = 2$

$6 - 2 = 4$

$18 - 9 = 9$

Answers to Worksheet #9

$10 - 2 = 8$

$7 - 6 = 1$

$1 - 1 = 0$

$10 - 9 = 1$

$13 - 5 = 8$

$7 - 0 = 7$

$12 - 4 = 8$

$10 - 8 = 2$

$13 - 7 = 6$

$12 - 3 = 9$

Answers to Worksheet #10

$17 - 8 = 9$

$9 - 7 = 2$

$6 - 5 = 1$

$2 - 0 = 2$

$8 - 4 = 4$

$2 - 1 = 1$

$8 - 3 = 5$

$3 - 2 = 1$

$8 - 6 = 2$

$9 - 9 = 0$

Made in the USA
Monee, IL
07 November 2026

68010600R00136